# Introducing Computerised
# Telephone Switchboards (PABXs)

## PUBLISHED BY NCC PUBLICATIONS

05799396

British Library Cataloguing in Publication Data

Introducing Computerised Telephone
Switchboards (PABXs)
1.  Telephone stations
I.  Title
621.385'02465          TK6211
ISBN 0-85012-364-X

First published in 1982 by:

NCC Publications, The National Computing Centre Limited,
Oxford Road, Manchester M1 7ED, England.

Typeset in 11pt Times Roman and printed by UPS Blackburn
Limited, 76-80 Northgate, Blackburn, Lancashire.

ISBN 0-85012-364-X

D
621·3858
INT

# Acknowledgements

NCC acknowledges with thanks the assistance of Mr P Lorentz of Telecommunications Management Ltd in the planning of the seminar upon which this book is based. We would also acknowledge the support of the Department of Industry's Computer Systems and Electronics Requirements Board.

# Acknowledgement

# Contents

*Note:* This book is based upon a seminar organised by the National Computing Centre in conjunction with the Telecommunications Managers Association, and run in Bristol, Manchester and London in May 1982.

The five presentations given at the seminar are reproduced here as four major chapters and a set of appendices; only minor editorial changes have been made to the original material.

# 1   PABXs – The State of the Art

P R D Scott (NCC)

## INTRODUCTION

Not so long ago, calculators were the size of typewriters and wristwatches had rotating hands. In the last decade progress in digital microelectronics has transformed both these products, and all the indications are that PABXs are about to undergo a similar transformation.

The Private Automatic Branch eXchange (PABX) provides a fully automatic internal telephone system, together with access to and from the public telephone network. Incoming calls are connected via the switchboard, and outgoing calls can be made either through the switchboard or automatically (usually by dialling 9 for an outside line). The new breed of computer controlled PABXs restores to an organisation the control and many of the features that existed in the old days of completely manual switchboards.

Computer-controlled or SPC (stored program control) exchanges are not new, but recently they have become digital instead of analogue. Digital technology offers the potential for carrying all forms of information – speech, text, data and pictures – in an integrated manner.

## THE TECHNOLOGY

### Analogue/Digital

A conventional analogue PABX converts audible speech into a continuously varying signal in the telephone handset. The signal is maintained in that form until it reaches the handset at the destination telephone where the electrical signals are converted back to audible speech.

In a digital PABX, speech is also converted into a continuous electrical signal, but this signal is then converted into a series of noughts and ones (called binary digits or bits) before switching. After switching, the string of bits is converted back to an electrical signal, and then back to speech at the destination telephone handset. The analogue to digital (a/d) conversion process usually takes place at the periphery of the PABX switch, but in the future, a/d converters are likely to be built into the telephone instrument, permitting fully digital operation over the extension circuit. The two common methods of analogue to digital conversion are known as pulse code modulation (PCM) and delta modulation.

In PCM, the electrical speech signal is sampled 8000 times a second, and the magnitude of the sample is assessed on a scale of $-127$ to $+127$. (The scale is non-linear according to a standard companding law, there being more points at the centre than at the extremities.) The magnitude of the sample is then coded as a series of 8 bits, called an octet. Thus each analogue speech signal gives rise to a digital signal of $8 \times 8000 = 64000$ bits per second.

In BT's trunk network, 30 PCM speech channels and two control channels are combined for transmission over a single circuit; the data rate on this circuit is $32 \times 64 = 2048$ kbit/s. Digital links of this rate are the basic building block of the UK digital network, and conform to the international CCITT standard.

In the USA, a different PCM structure is used; known as the T1 system it operates at 1.544 kbit/s and is not compatible with the European system.

Delta modulation also involves sampling the analogue speech signal, but at a very fast rate of 50,000 bits per second or more. In this case it is the *difference* between one sample and the previous one which is coded as either a 0 or a 1 depending on whether the sampled signal has decreased or increased in magnitude.

### Stored Program Control (SPC)

An SPC exchange is computer controlled: the behaviour of the exchange is governed by a program stored in the exchange computer's memory. This not only makes the exchange capable of providing many sophisticated facilities, but also means that these facilities can be tailored to meet the needs of individual users, and altered quickly and easily by amending the computer program. Extension numbers, for example, can be altered at will.

This contrasts with the non-SPC exchange whose operation is governed by the exchange wiring. Scope for providing additional facilities is limited and even the simplest alteration requires rewiring.

The use of SPC is unrelated to whether a PABX is analogue or digital, but in practice all digital PABXs are SPC exchanges.

## SWITCHING

In a conventional PABX, each call is allocated a two-wire path through the exchange; the number of simultaneous calls possible depends on the number of switches provided. The technique is sometimes referred to as space division switching – to differentiate it from the newer technique of time division switching used in digital exchanges.

Over the years, PABXs have employed a variety of switching systems, ranging from Strowger switches, to crossbar, reed relay, and solid state electronic switches which have no moving parts.

### TIME DIVISION SWITCHING (Figure 1)

When a speech signal has been converted into a series of 0 and 1 bits, it becomes possible to carry many conversations on a single high-speed circuit by interleaving bits from different conversations, a process known as time division multiplexing. The reverse process is known as demultiplexing and permits the individual bit strings to be retrieved.

Modifying this process by the addition of a memory turns time division multiplexing into time division switching. Interleaved pulses on the high-speed circuit are read in order into a set of memory stores. The stores are then emptied in a different order, their contents being fed out on another high-speed circuit. The effect is to switch the individual conversations from one channel to another, the routeing being governed by the order in which the

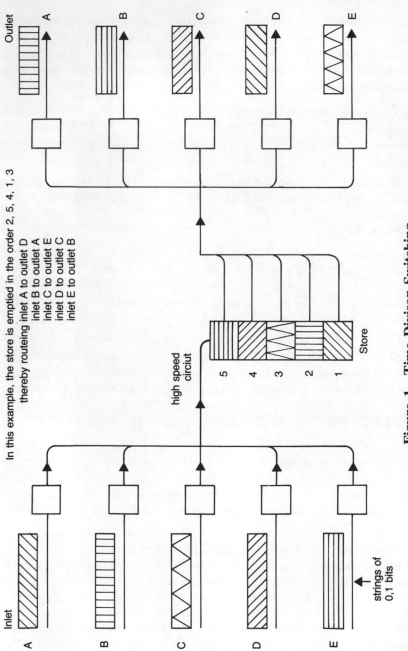

In this example, the store is emptied in the order 2, 5, 4, 1, 3
thereby routeing inlet A to outlet D
                 inlet B to outlet A
                 inlet C to outlet E
                 inlet D to outlet C
                 inlet E to outlet B

Outlet

A

B

C

D

E

high speed
circiut

Store

5
4
3
2
1

Inlet

A

B

C

D

E

strings of
0,1 bits

**Figure 1   Time Divison Switching**

memory stores are emptied. Thus there is only one path through a time division switch and this is shared in turn by elements of each call in progress. The two directions of transmission (A to B and B to A) are, however, handled separately providing '4 wire switching'.

## ANALOGUE AND DIGITAL PABXs

The future certainly lies with digital SPC PABXs, if only because they will inevitably become cheaper to manufacture than analogue exchanges (Figure 2).

That is not to say that a digital PABX is necessarily the best buy for an organisation replacing a PABX today; a cheaper SPC analogue PABX can provide the same facilities as an SPC digital exchange. So what is the justification for paying extra for a digital PABX? Three reasons in favour of the digital option are:

— *The PABX will be in a city centre location likely to have access to a local System X exchange within the next five years.*

BT's programme for modernising the UK telephone network spans a period of thirty years or more, but most of the larger city centres are likely to have System X exchanges installed within the next five years. The following list (published by BT in February 1981) shows when most of Britain's towns and cities will be getting their first System X exchange, but naturally, a massive project involving the replacement of the entire UK telecommunications network takes time and money; completion won't be before the end of this century.

| CENTRE | OPEN DATE | CENTRE | OPEN DATE |
|---|---|---|---|
| Aberdeen | 1986 | Baynard House | |
| Ayr | 1985 | (London) | 1984 |
| Aylesbury | 1985 | Bishop Auckland | 1987 |
| Basildon | 1985 | Blackburn | 1987 |
| Bradford | 1987 | Blackpool | 1986 |
| Belfast | 1986 | Birmingham | 1983 |
| Bedford | 1987 | Bournemouth | 1986 |
| Bathgate | 1985 | Brighton | 1987 |

16

TELEPHONE SWITCHBOARDS

| CENTRE | OPEN DATE | CENTRE | OPEN DATE |
|---|---|---|---|
| Bristol | 1984 | Manchester | 1983 |
| Bishops Stortford | 1984 | Northallerton | 1987 |
| Cambridge | 1984 | Nottingham | 1985 |
| Cardiff | 1986 | Northampton | 1985 |
| Chester | 1987 | Newport | 1987 |
| Colwyn Bay | 1986 | Newcastle | 1985 |
| Colchester | 1987 | Oxford | 1987 |
| Coventry | 1983 | Potters Bar | 1987 |
| Darlington | 1987 | Peterborough | 1987 |
| Derby | 1987 | Plymouth | 1986 |
| Dundee | 1987 | Portadown | 1987 |
| Dudley | 1986 | Preston | 1986 |
| Edinburgh | 1984 | Portsmouth | 1987 |
| Exeter | 1985 | Reading | 1985 |
| Guildford | 1983 | Redhill | 1987 |
| Greenock | 1987 | Rhyl | 1987 |
| Glasgow | 1984 | Southend | 1985 |
| Hexham | 1985 | Sheffield | 1986 |
| Hcniton | 1987 | Skipton | 1984 |
| Haverfordwest | 1987 | Slough | 1984 |
| High Wycombe | 1987 | Southampton | 1985 |
| Ipswich | 1985 | Swansea | 1985 |
| Inverness | 1986 | Stoke-on-Trent | 1986 |
| Lancaster | 1987 | Swindon | 1987 |
| Londonderry | 1987 | Shrewsbury | 1986 |
| Leicester | 1984 | Thetford | 1987 |
| Leeds | 1983 | Tunbridge Wells | 1986 |
| Luton | 1985 | Warrington | 1986 |
| Liverpool | 1984 | Weybridge | 1986 |
| Maidstone | 1986 | Whitby | 1986 |
| Medway | 1987 | Wigan | 1987 |
| Maidenhead | 1987 | Wolverhampton | 1987 |
| Middlesbrough | 1985 | Worcester | 1987 |
| Monmouth | 1987 | York | 1987 |
| Morpeth | 1985 | | |

Note that some of these dates have since changed. The up-to-date situation should be checked with BT System X Marketing Group, ME/RCS2.2.1, 2nd Floor, Seal House, 1 Swan Lane, London EC4R 3TH. 01-357-2899.

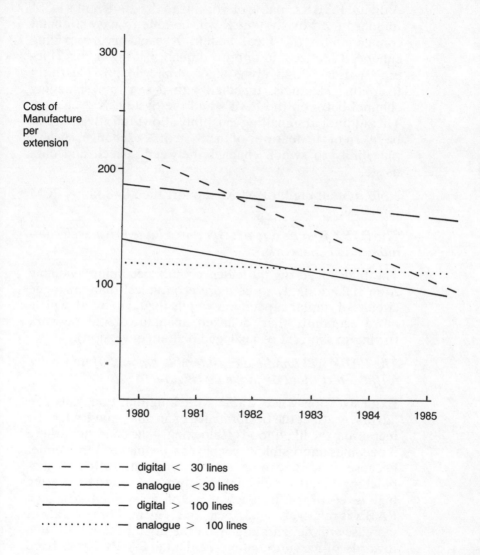

**Figure 2   PABX Manufacturing Costs
Analogue/digital, 1980/85**

A digital PABX capable of working to the 2048 kbit/s PCM standard used by System X will be able to have a digital connection to the local System X exchange, providing enhanced service and improved quality at lower cost (Figure 3). Each 2048 kbit/s digital link will provide thirty telephone channels, together with a separate signalling channel between the PABX and the System X exchange. This additional signalling capability allows full advantage to be taken of the features of the System X exchange, such as the ability to switch channels between speech and data usage.

Note: not all digital PABXs support the 2048 kbit/s PCM interface.

— *The PABX is to be one of several in a large private telecommunications network.*

The high-speed digital leased circuits becoming available from BT are likely to be more economical than analogue circuits of similar capacity. A wholly digital network will be more economic than a mixed analogue/digital network (owing to the cost of analogue/digital conversion).

— *The PABX will handle a considerable amount of non-voice traffic – text, data or facsimile (Figure 4).*

Extensive non-voice traffic with long-duration calls can adversely affect the performance of an analogue PABX, by increasing the likelihood of 'blocking' – the condition where it becomes impossible to connect a free inlet to a free outlet because no route can be found through the switch. A non-blocking digital switch is inherently better suited to carrying high levels of traffic. The volume of non-voice traffic on a PABX is relatively small today but is increasing; the rate of increase will depend largely on the organisation's attitude towards office automation. A digital PABX has a fixed maximum data rate – typically about 64 kbit/s, which is ample for most data applications. An analogue PABX imposes no such fixed limit.

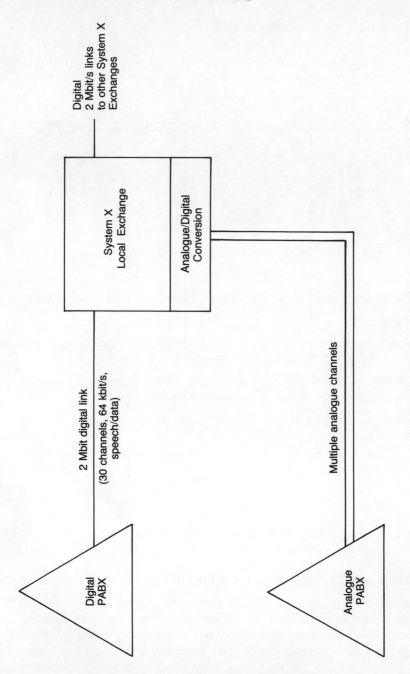

Figure 3  System X

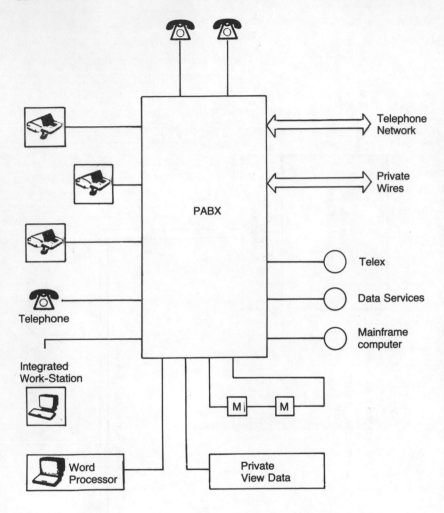

**Figure 4    System Architecture**

## PABXs AND LOCAL AREA NETWORKS

Most local area networks developed in the computing world to provide communications between terminals and computing resources on the same site. These networks filled a gap between the high-speed short distance data highway over which a mainframe computer communicates with its peripherals, and the low-speed long distance data networks providing terminal access to a computer centre over BT circuits (techniques were borrowed from both).

Local area networks usually have a single cable of some sort to which all devices are attached. Typically, the cable is up to 1000 metres long, can support several hundred devices, and transmits data at several megabits per second (Figure 5). Twisted wire pairs, multicore cable, coaxial cable and optical fibre are all used. Messages are broadcast on the network in the form of packets carrying the address of the destination device. All devices on the network are aware of a packet being sent, but only the device to which the packet is addressed actually reads the message. Various rules are imposed to ensure that all devices receive fair treatment, and that messages do not collide and become garbled.

Local area networks are being installed to provide services such as electronic mail, access to on-site data processing facilities, and word processing.

It is possible to handle voice traffic on a local area network, but the technology will have to prove that it provides substantial benefits if it is ever to usurp the PABX as the prime system for voice communications.

In the short-to-medium term, the trend seems to be in the direction of a hybrid local area network/PABX approach, with the PABX appearing as a gateway on the local area network and providing access to and from the outside world.

## LIBERALISATION

The British Telecommunications Act of 1981 removed BT's monopoly on the supply of equipment for connection to the public telephone network, but its measures do not become fully effective until July 1983, three years after Sir Keith Joseph's initial announcement that the monopoly was to be lifted.

The three-year period was part of a deliberate phasing-in period to give British industry time to put products on the market to compete with imports from the USA and elsewhere. It also gave time to draw up the necessary technical standards to which equip-

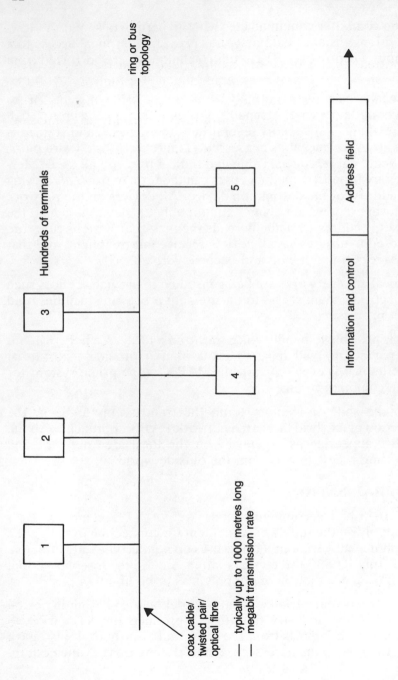

**Figure 5   Local Area Networks**

ment connected to BT's networks must conform.

The Act itself is not specific in terms of what is and what is not permitted. It is a piece of enabling legislation placing the real power in the hands of the Secretary of State for Industry. As far as PABXs are concerned, the situation is as follows:

— The British Standards Institution is drawing up a standard for PABXs, to be published in July 1983. Once the standard is published, any supplier will be able to submit a PABX to the British Approval Board for Telecommunications for evaluation against this standard. Only models which conform to the standard will be permitted connection to the BT network;

— Until the PABX standard is published and PABXs are 'liberalised', the existing situation prevails: ie BT has a monopoly on the supply, installation and maintenance of PABXs of less than 100 lines, and on the associated extension telephones and wiring ('block wiring'); above 100 extensions BT has approved certain PABXs for private supply and installation, but retains the monopoly on the supply and installation of extension telephones and wiring, and on the maintenance of the whole system;

— After July 1983, all PABXs, together with extension instruments and wiring, will be open to competitive supply and installation. BT will continue to do the commissioning and acceptance testing of all PABXs, and to approve the wiring;

— Non-BT maintenance will be permitted on time division SPC PABXs installed after July 1983. Other types of PABX and *all* PABXs installed before July 1983 will continue to be maintained by BT;

— To improve the choice of large PABXs available to users before July 1983, six PABXs have been selected by the Department of Industry in consultation with BT, to undergo an accelerated BT approval procedure. This move was announced in January 1982 and approval is expected to take about a year.

Figure 6 summarises the situation before and after July 1983.

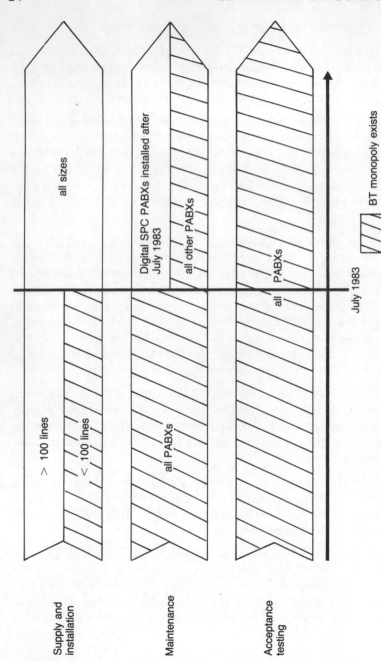

**Figure 6   PABX Liberalisation**

## Summary of Liberalisation Stages

One of the reasons for 'liberalisation' was to encourage innovation in telecommunications products and services, and considerable competition is expected in the PABX market, particularly at the smaller end (less than 100 extensions). There is likely to be a sudden expansion in the number of suppliers, followed by a contraction as the competition becomes fiercer with several suppliers going out of business. Buyers need to assess the stability of the company and not just the product being offered. Profitability, size and structure of the organisation, and research and development activities all provide some guide to the stability of the supplier.

## CONCLUSION

There is already considerable activity in the PABX market, but there is more to come. The advice offered to anyone currently considering replacing a PABX is to wait, if possible. Digital SPC PABXs are becoming more economic and after July 1983:

— there will be more choice (but check out the supplier as well as the product);

— non-BT maintenance of digital SPC PABXs will be permitted.

# 2 PABX Facilities, Features and Services

J B Waites
(Telecommunications Managers Association)

## INTRODUCTION

Today telecommunications is assuming a new and growing importance in a changing society. The current information revolution associated with microelectronics has led to the general belief that efficient, effective telecommunications systems are required to move large quantities of information rapidly. In future, there will be a growing demand for video information services, in addition to the telephone, as a means of exchanging information.

## THE SPC PABX

An essential element of the business communications systems envisaged will be the electronic stored program controlled PABX which is now being installed in ever increasing numbers.

In simple terms, the electronic SPC PABX is no different from any other PABX in that:

— it has switchboards (albeit small, desk top, and modern);

— it has an equipment room (although this is much smaller than its PABX 3, etc counterparts);

— it has lines and extension telephones which can be connected together to make calls.

However, there are one or two distinct advantages:

— It has many additional user facilities when compared to the original PABXs, and many of these facilities are extremely sophisticated;

27

— Being electronic, as opposed to electromechanical, it has virtually no moving parts and is thus much more reliable and less susceptible to wear. Its advanced design also gives it great potential for carrying the 'office automation' information traffic of the future;

— Management information and control over the use of the system is much more sophisticated; thus, the electronic SPC PABX will allow companies to manage their telecomms services more efficiently;

— SPC systems have the potential to interconnect business terminals in a wide variety of shapes and sizes and applications.

A brief summary of how PABXs have developed from manual telephone systems follows:

— the original manual switchboard (the PBX) connected *all* calls via a human operator;

— the first generation PABXs (electromechanical systems like the PABX 3 and PABX 7) gave much more freedom of action to the user, but at the expense of the company;

— the new generation SPC PABX provides the user with most of the facilities originally provided by the manual switchboard, but under much tighter control.

## THE UK SCENE

Present policy in the UK requires that British Telecom must handle supply and maintenance for PABXs of less than one hundred extensions capacity. However, PABXs of greater capacity are purchased from one of a small number of BT approved PABX suppliers and are subsequently maintained by BT to protect its maintenance monopoly. BT has no direct control over the design of new PABXs that an approved supplier may wish to market, but initially, the number of PABXs installed is limited until BT confirms that certain performance and maintenance standards are met.

With deregulation the situation is changing. Future PABXs will have to meet BSI standards and be approved by the BABT. After

July 1983, the maintenance of digital SPC PABXs will be thrown open to any approved contractor. Until 1978, the majority of PABX suppliers sold electromechanical PABXs, either the Strowger, eg PABX 3 and PABX 4, or latterly, the Crossbar type, eg ARD 561/2. These two types utilise technology developed some sixty years ago and, although there has been considerable development in the components used during this period, the telephone switching principles employed have stayed the same. IBM was the one exception to this rule having introduced the very successful 3750 SPC PABX in 1973.

Between 1978 and 1980, each PABX supplier announced the intention to market a completely new technology range of PABXs, known as stored program control (SPC) PABXs. Each SPC PABX is controlled by fully duplicated microprocessors or computers which instruct the switching matrix to set up the required connection. SPC PABXs utilise solid state or electronic switching. Furthermore some SPC PABXs use analogue techniques within the PABX whilst others have opted for digital switching. In both cases, the external interfaces of any SPC PABX (ie for connection to extension telephones, BT's public network or to other PABXs) are standard, because they have to meet stringent BT specifications.

However, it could be said that digital SPC systems of a type which will support European digital transmission standards, ie 2 megabit thirty channel links, will be in a much better position to service the office communications requirements of the future. (This envisages the widespread use of text communications terminals for message transmission and computer access.)

There are currently eight SPC PABXs available in the UK, including two small systems which can be leased from BT: GEC SL1; IBM 3750; IBM 1750; ITT 4080; Plessey PDX (also marketed by Telephone Rentals); Philips EBX 8000; BT Monarch; and BT Regent. The SL1 is supplied by Reliance Systems Ltd in the UK and the 'UK manufacturer' is GEC. In fact, the PABX is designed and manufactured by Northern Telecom in Canada with only the UK interfaces designed and manufactured by GEC. The SL1 is digital, and has a maximum capacity of over 2000 extensions.

IBM currently manufactures and supplies two SPC PABXs, the 3750 and 1750, both made by IBM France, both employing analogue switching techniques. The IBM 3750 can handle up to 2500 extensions, whilst the more modern 1750 is aimed at the market below about 750 extensions.

The ITT 4080 is supplied by ITT Business Systems and manufactured by ITT Germany where it was mainly developed. A significant amount of software development was performed in the UK. The maximum capacity of this analogue PABX is approximately 2400 extensions.

The PDX is offered in the UK by two suppliers, Plessey Communications Systems and Telephone Rentals, and manufactured by Plessey Telecommunications Ltd. The original PDX 800 (with a maximum capacity of 800 extensions) was designed and developed by the Rolm Corporation of California and the early systems were manufactured in the USA. During 1978 Plessey Telecommunications Ltd set up a production/assembly line in the UK but it is understood that a significant proportion of the hardware is still imported from the USA. The PDX 800 was fully approved in late 1979 and there is now a customer base of well over 100 systems in service. Plans exist to extend the capacity of the PDX up to 2000 lines. The PDX is a digital PABX.

The EBX is supplied by Pye Business Communications and manufactured by Philips Telecommunications, Holland. The design and development was mainly performed by Philips, with some hardware and software development for special UK requirements by Pye. The maximum capacity of this analogue PABX is 8000 extensions. Reed relays are extensively used for switching.

BT offers the Monarch and the Regent SPC systems, both with a maximum capacity of around 120 extensions. The Regent is an analogue system whilst the Monarch is digital. With the Monarch 120C, BT plans to expand the system up to 240 extensions and offer a CCITT two megabit digital interface.

Further systems are believed to be under evaluation including: Thorn Ericsson MD 110; Mitel SX 2000 and ICL DNX 2000 (Mitel); and GTE Ferranti GTD 1000.

Entering much later into the SPC PABX market, the MD 110 will be supplied by Thorn Ericsson and manufactured by L M Ericsson of Sweden. The design and development is a joint enterprise between both companies. Currently it seems that the MD 110 will have a capacity of up to 10,000 extensions. It is a digital switching system employing distributed processing techniques and is designed to be compatible with System X digital methods and CCITT two megabit link standards. The Mitel SX 2000 was announced in the UK in 1980. This is a fully digital system with a CCITT two megabit interface and digital extension 'phones. It has a maximum capacity of 10,000 lines and is currently undergoing BT evaluation. The first systems are expected to be in service and on trial by the end of this year. The SX 2000 is also to be sold by ICL as their DNX 2000. It is anticipated that the DNX 2000 will be enhanced to carry many forms of user based text and data traffic including interworking with computer based information systems.

GTE is a recent entrant to the UK PABX market having made a trading agreement with Ferranti (UK) about a year ago. The GTD 1000 SPC PABX is digital and capable of handling up to 1000 extensions.

## FEATURES AND FACILITIES

Let's now take a look at the features and facilities available to users. The terminology used tends to vary from supplier to supplier: here, the aim is to apply the most commonly used terms to describe some of the most popular facilities.

### The Telephone Instrument

The first and most obvious feature facing the user will be the' telephone instrument. This is likely to be a pushbutton keyphone, replacing the traditional rotary dial equipped telephone, which enables the user to take advantage of the wide range of facilities and faster calling potential of SPC.

The keyphone presentation is standard as shown below. (The keys marked * and #!are used for initiating and closing instructions to the exchange.)

**Figure 7**

**User Facilities**

1.  Abbreviated dialling:

    Normal telephone number (containing typically 10 digits for
    an STD call, 14 for an overseas call) replaced by 2 digit code.

    —  easier to remember;

    —  easier to use;

    —  quicker;

    —  reduces misdialling problems.

    Typically 20-100 numbers can be stored.

    Can be 'system' short code list or individual (subject to 'bar-
    ring' constraints).

2.  Call back (ring when free):

    Called extension engaged, caller keys two digit code and
    replaces the handset;

    System rings back when both parties are free;

    Can also be used on 'no reply' – the PABX detects when the
    called party returns by resumption of telephone activity.

3.  Call diversion:

All calls ('follow me'): – User keys a code and extension number visiting.

—  cancel from original or secondary extension.

On no reply: – call diverts to selected alternative extension after a predetermined period of ringing, ie ten seconds.

—  normal outgoing facilities retained;

—  useful for customer service phones, absent executive.

On busy: – diverts to nominated alternative immediately the busy condition is detected.

—  retains full O/G;

—  useful for customer service, alternative answering.

4.  Call pick-up:

Directed: any extension can pick-up/answer calls to another extension – user dials pick-up code plus extension number.

Group pick-up: specified extensions in pick-up code only; extension can be number of only one pick-up group.

5.  Call restriction:

Unrestricted: – call anywhere;

Trunk restricted: – local calls only; cannot dial (Key) STD calls ('0');

Semi restricted – eg all IDD (010) barred;

Fully restricted – all outside calls barred plus selected barring, eg '192', '123', etc;

System abbreviated codes allowable.

6.  Conference:

Multi party – number of extensions on one conference varies between systems (eg PDX has 8) but never more than one external line.

7.  Last number repetition:

    Automatic – key in code at any time;

    Selected – key in 'store' code whilst call attempt held – key repeat code later.

8.  Secretarial:

    Replace Plan 107;

    Incoming call to manager diverts to secretary, who – keys manager by single digit, offers, then transfers. Outgoing service as normal both phones – can be changed at holiday times – secretary's extension can handle more than one manager extension.

    As you will have gathered, the use of these facilities requires the user to remember and use many different codes. Such a system is not particularly user friendly. The alternative would be to install 'featurephones' which present the more popular facility in a single press button fully labelled with the feature it controls. Such telephones exist and will be offered in the not too distant future.

### System Facilities

1.  Direct inward dialling:

    Each extension so equipped directly accessible from BT public network;

    Requires separate number on a main BT exchange;

    Incoming calls can then be transferred;

    Constrained by BT ability to provide exchange lines (numbers).

2.  Hunting:

    Number of extensions associated to enable incoming calls to be offered around the group.

    Special 'group' extension number allocated.

    — cyclic – incoming calls shared equally;

— sequential – call offered to first free extension in sequential numbering scheme.

Useful in general department where all dealing with some business, eg service desk.

Individual extension numbers retain normal characteristics.

3. Night service:

Selected – designated extensions directly connected to exchange lines – inflexible;

Level 8 – any extension – full transfer capability – night service bells.

4. Call Logging:

In today's world we cannot let a discussion about PABXs pass without mentioning call logging, or CILE to the trade. Briefly, CILE allows the PABX manager to know who dials what and when, and to process how much it costs. In practice, it is more convenient to report in detail only on the high cost items, ie normally the PABX manager is told in detail about each expensive international call, but only sees the bulk cost figures for local calls.

The original CILE produced all the call information onto industry compatible magnetic tapes for computer processing, and the printed management information was frequently old when received. More recent CILE has a real time reporting facility for reporting on serious exceptions as and when they happen, and many CILE systems now contain a built-in data processing system so that up-to-date management information can be produced very quickly.

SPC PABX manufacturers have been pretty vague to date on what CILE facilities their systems provide. From a naive belief that everything will be possible, it now seems that most systems provide little more than an industry compatible interface to enable all call data to be written to magnetic tape for subsequent processing. Some systems provide a modicum of traffic data including total dialled units per extension (mainly for accounting purposes). By and large you still need to add in

some CILE equipment, though with the refined service class-
ifications now available this may be a less pressing considera-
tion.

## MANAGEMENT IMPLICATIONS

Having looked at what the extension user can expect from SPC,
what about management?

Because these SPC PABXs are designed with standard external
interfaces, the different methods of working within the PABX are
of no importance to the average telephone user. The most important benefit of the SPC PABX is that it is computer and therefore
program controlled and this provides flexibility in three main
areas:

1. A large number of user facilities can be provided by software means. Some of these software facilities replace existing facilities provided by hardware, such as special telephone arrangements, like the Plan 107. Others are completely new facilities.

2. The allocation of an extension number to a particular telephone can be changed by a user representative (normally
   the telephone supervisor) using a teleprinter with access to
   the PABX program. For example, if a user moves to a new
   office, the existing extension number and telephone
   facilities can be transferred immediately without the delay
   and costs incurred if BT carries out the work. It is estimated
   that approximately 70% of all telephone moves and changes
   currently carried out by BT could be performed by a user
   representative on an SPC PABX and lead to significant
   savings.

3. New facilities, or improvements to the SPC PABX, can be
   installed quickly without affecting telephone service by
   loading a new software release. This is identical to the
   computer practice of improving system efficiency and
   facilities by the regular updating of software. In this way, an
   SPC PABX can be considered to be 'future proof' until a
   further dramatic change in technology occurs.

## MANAGEMENT BENEFITS

There are a number of other benefits apart from improved user facilities, such as:

— Less PABX equipment accommodation is required. In practical terms, on a 500 line SPC PABX you could be saving 800 sq ft;

— Faster delivery and installation are possible. You get it when *you* want.it, and you can subsequently move it very quickly if you need to.

### Cost Savings

From a cost point of view there are a number of areas which could be seen as offering cost saving opportunities:

— The totally flexible barring of outgoing calls can provide savings in BT call charges without adversely affecting staff's working efficiency. PABX suppliers suggest that a 10% saving is possible by judicious use of this facility;

— Savings in accommodation costs for a smaller PABX equipment room. Typically, a 500 line SPC PABX saving 800 sq ft corresponds to savings of £15,000 p.a. in London office rent terms;

— Savings in connection charges and rental for special BT telephone instruments which are not required on SPC PABXs (eg Plan 107), although this could be somewhat offset by the increased rental for keyphones which *are* required;

— A reduction in BT charges for telephone moves and changes, as the majority can now be performed by the telephone supervisor;

— An unquantifiable improvement in staff efficiency brought about by the effective use of the SPC PABX facilities.

## CONCLUSION

Three main notes of caution are worth considering:

— Many of your users will resist changes and will not bother to use any of the new and sophisticated facilities. They should be encouraged to do so if you wish to get the best from your new system. Good publicity and training are essential;

— Your telephone operators are also liable to be unenthusiastic about SPC, partly because of a natural resistance to change and partly because SPC systems tend to reduce switchboard operating duties to an absolute minimum. You must involve the operator more in system maintenance work and other duties, but switchboard operating is certainly very different and potentially much less satisfying than in the old 'cord and lamp' days;

— The erosion of BT control over PABX standards places more and more onus on the customer getting it right. It really is now a situation where one cannot say too loudly 'let the buyer beware'.

# 3 Replacing Strowger by a Stored Program Control PABX

K R Clark (BL Systems Ltd)

## INTRODUCTION

The majority of telephone users today are familiar with using electromechanical telephone exchanges with their own domestic telephone at home. In the business environment, the majority of people still rely on electromechanical telephone exchanges, whether fully automatic or manual, to make calls within their own location or company or onto the Public Switch Telephone Network (PSTN).

At some point in time you will reach the stage where the replacement of your own company's telephone exchange must be considered. It could be for many different reasons including increased capacity, the need for additional facilities, reliability/age factor, rebuilding or changing offices, economy measures, or removal to new accommodation with the need to provide completely new telephone exchange facilities.

Within my own company, BL Systems Ltd., we have met these situations many times in the past, but during the last three to four years we have considered installing the new generation of SPC PABXs. We have in fact installed a number of SPC PABXs for BL sites, such as the EBX8000 from Pye, IBM's 3750 and the Reliance SL1.

This chapter will be useful to the communications office manager who is considering the purchase of a new SPC PABX exchange, and needs to be more aware of the planning and implementation details. BL's experience of a particular SPC

PABX installation may be of assistance, particularly at the early planning stages, and will identify many aspects that must be taken into account when considering SPC PABXs.

The main areas to be considered are: planning; justification for a new exchange; factors governing the choice of a digital PABX; and implementation. Technical descriptions, details of facilities, training or PABX maintenance requirements are not discussed.

## BACKGROUND

In the late 1970s, a new Vehicle Engineering and Test Facility site was opened on a green field site in Warwickshire. A PABX 7 was rented from British Telecom to provide telephone facilities for the site and was inter-connected onto BL Systems' Internal Telephone Network via microwave radio links (Figure 8), which allowed access to over 50 other BL locations.

As the site was developed and the number of people who needed telephone facilities increased, heavy demand was placed on the PABX 7. This resulted in exchange congestion together with a shortage of telephone extensions. The PABX 7 had 10 exchange lines, 10 private wires and a maximum of 90 extensions. It is not possible to extend the capacity of a PABX 7, therefore the only option was to change to a larger size PABX. A project was set up under the control of a project manager to consider the options.

## PLANNING

It is important to mention at this stage the need to appoint a project manager to coordinate the many aspects involved in providing or replacing a new PABX telephone exchange.

Consider the number of parties involved and the various aspects of what may initially appear to be a fairly straightforward task:

— The Company Project Manager for PABX;

— Site facilities;

— Site services;

— Systems data personnel if involved;

— Finance;

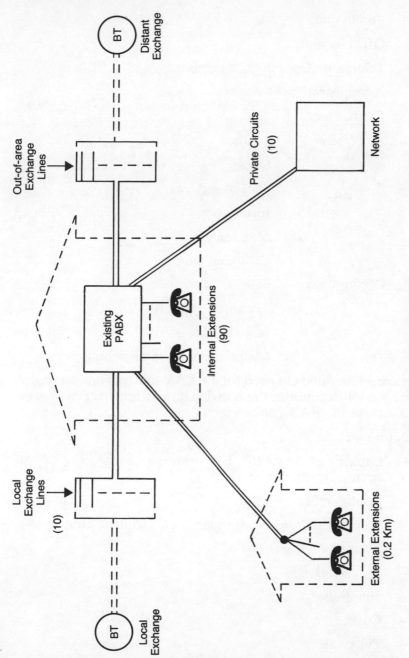

**Figure 8   PABX 7**

— Purchasing;

— Office layout;

— Telephone Operators/Supervisor;

— Administration Managers;

— Contractors.

British Telecom    –  Traffic

                      Sales

                      Special services

                      Installation

                      Maintenance

                      Customer support

PABX Suppliers    –  Sales

                      Technical support

                      Installation

                      Maintenance/Commissioning

Having identified the need for a PABX and appointed a project manager, let us consider the individual tasks undertaken in order to complete the PABX installation.

Determine:

— Capacity of the PABX. Extensions, exchange lines, private circuits, traffic handling, etc;

— Configuration;

— Future requirements (growth, new facilities, etc);

— Type of PABX;

— Specification;

— Suppliers;

— Cost;

— Timescales;

— Accommodation;

— Facilities required;

— Justification;

— Finance;

— Order;

— Delivery;

— Installation and commissioning details;

— Telephone instruments;

— Class of Service allocations;

— Maintenance;

— Training needs;

— Operational services;

— Monitoring/Records/Documentation.

Planning any PABX installation is important but even more so when *changing* a telephone installation. Under no circumstances should a business customer be left without telephone service.

Before the choice of PABX can be made, a full specification of requirements for the PABX must be compiled. A traffic survey will be carried out by British Telecom – this indicates the quantity of lines and equipment that will be needed (Form A5245). A reasonable period of notice is required by BT to perform this survey. From this traffic data form, the capacity of the PABX can be determined. In BL's case, the specification for the PABX exchange capacity was as follows:

| | | |
|---|---|---|
| Exchange Line | 25 | |
| Extensions | 248 | |
| Operator's Positions | 2 | |
| PWs | 24 | (E & M Signalling) |

## CHOICE OF PABX

Once the basic specifications had been prepared, the next step was to consider the types of PABXs available, the various suppliers concerned and the constraints that affect the choice of PABX. The latter include specification, cost, timescale and accommodation. The prime factor is specification, and in BL's case, a PABX of 248 extensions was identified. BT is unable to supply an exchange of this capacity, therefore, private suppliers were the only option.

Now to consider the possibility of a digital PABX. Today it is less likely that a customer will install a Strowger type electromechanical PABX for some of the following reasons:

— Strowger exchanges are obsolescent;

— They are unlikely to be supported by spares and extension modules in the 1990s;

— Experienced maintenance labour will become scarce and expensive as Strowger equipment is phased out;

— Strowger does not offer the facilities or flexibility available on SPC exchanges;

— Will not be compatible with System X being installed by BT;

— Larger accommodation required;

— Longer and more involved installation;

— Becoming more expensive than SPC exchanges;

— Costs for Strowger equipment have increased over the last two years while those for SPC exchanges have reduced.

Indeed, today many manufacturers will not offer a Strowger PABX exchange to customers, and when they do so, it is unlikely to cost less than an SPC PABX. Electromechanical Crossbar PABXs have similar constraints and will not be considered further.

In conclusion, it is not worthwhile considering Strowger or Crossbar PABXs unless specific reasons apply, eg compatibility with existing equipment.

**Digital or Analogue**

Taking SPC exchanges a stage further, on what basis should we decide upon digital or analogue PABXs? Some may argue that currently there is very little to choose between either.

From the telephone extension user's point of view, little difference is seen. The main benefits lie in the switching methods where a digital PABX may be designed around a time division non-blocking switch. This allows greater traffic capacity than analogue PABXs. In practice however, the traffic capacity is generally limited by network design and does not necessarily offer a significant advantage over analogue. Digital transmission of speech is less prone to noise and crosstalk. The trend is towards sending and receiving data via digital circuits. The ability to switch digital lines must therefore be considered when connection onto the PABX is involved.

In addition to the reasons mentioned above, the main impetus for selecting digital SPC PABXs at BL Systems is to achieve compatibility for the future.

Most digital SPC PABXs should be able to achieve compatibility with System X from BT, and with digital networks which could be planned within a company. Therefore, we would recommend installing a PABX designed to achieve compatibility with future transmission systems and office technology. The choice between a digital or an analogue SPC PABX may not really be significant when other factors and constraints relating to the purchase are considered, eg price, delivery, etc.

## SELECTION

Having decided upon a digital PABX, we looked at the choice of suppliers. Only two types of digital PABX met British Telecom (then the Post Office) approval, the SL1 from Reliance Systems Ltd., and the PDX from Telephone Rentals Ltd. and Plessey Communications Systems Ltd. Quotations were obtained and compared for the two systems:

| | |
|---|---|
| — *Cost* | — There was no substantial difference in cost. The cost per telephone line was approximately £840; |

| — *Delivery* | — | Both PABXs would be available within our 6 months delivery requirement; |
| — *Installation* | — | A further period of 2 months was needed for installation and commissioning for both systems; |
| — *Facilities* | — | No significant difference; |
| — *Accommodation* | — | The SL1 did not need rear access to the equipment cabinets, otherwise both PABXs need standard accommodation only; |
| — *Technical Specification* | — | There was some difference in specification between the PDX and SL1; |
| — *Capacity* | — | Both PABXs had growth up to 1,000 lines. |

**Our Choice**

Our choice for a digital PABX was for a 220-line SL1 from Reliance Systems Ltd. The reasons for deciding upon the SL1 were:

— the benefits obtained in accommodating the PABX;

— lower power consumption;

— less cost involved in supplementary equipment;

— better flexibility within the PABX;

— software problems that other customers had experienced with PDX.

**JUSTIFICATION**

The justification to purchase a PABX, based on the foregoing analysis, was submitted to our Financial Department. The options were:

— Do nothing. This alternative was rejected since the existing exchange was already overloaded and inadequate;

— Electromechanical PABX. Initially, this alternative would be the cheapest option but would not meet the accommodation requirements, would not be compatible with future systems and costs would increase in the future;

— Digital PABX, as recommended in the proposal.

The proposal involved replacing the existing PABX 7 with a Reliance SL1 PABX electronic based exchange capable of easy extension, if required, to meet further site development (Figure 9).

The stored program control PABX offers many operational benefits:

— The systems may be purchased and installed easily in modular form to meet site requirements;

— It can handle from 100 to 1,000 telephone extensions;

— It is more compact and considerably quieter than the existing type of exchange;

— The additional space required to accommodate the enlarged capacity exchange can be found within the existing room;

— It may be programmed to operate as a separate exchange for other tenants, if so desired;

— It will be compatible with the new British Telecom System X;

— It offers a long list of standard facilities to users, telephone operators and exchange administrators;

— Replacement parts and service will be available for the expected life cycle;

— Internal telephone numbers may be changed by reprogramming, instead of requiring the attendance of a British Telecom engineer to physically alter the wiring;

— It is self monitoring for performance, call analysis and traffic analysis;

— No special working environment is required;

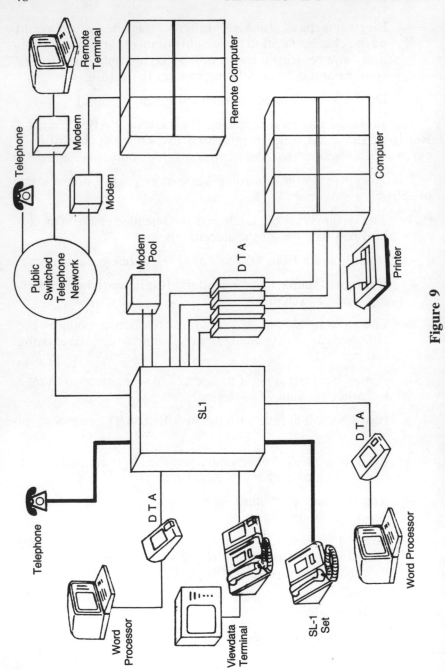

**Figure 9**

— It represents the latest technology available in general service;

— Offers management control over site facilities.

British Telecom connection costs for the PABX and telephone facilities amounted to £14,000, in addition to the capital cost of the PABX equipment.

## OTHER FACILITIES

It was decided to include additional facilities for the BL Systems' PABX installation:

1. Direct Dialling In (DDI) – (Not DID – a PABX feature). We were able to include DDI on the PABX because it became available on the local British Telecom PSTN exchange. The main benefit of DDI, in addition to the convenience that this facility offers, was that only one operator's console position would be needed. 150 DDI extensions were ordered from British Telecom at a cost of £2,200. Line cards on the PABX had to be substituted but no additional cost was involved. Stand-by battery capacity had to be increased from 4 hours to 6 hours. Savings were made on the one operator's console.

2. Call logging (CDR 1000): The amount of call logging and telephone traffic information is limited unless the full CDR 1000 Telephone Management and Call Information Logging System is included with the PABX (meter pulses were requested on the 25 exchange lines for this facility).

3. Halon Gas Protection: At the request of our fire safety officer, a BT approved Halon Gas Protection system was installed to protect the PABX. The Halon system cost approximately £2,300. British Telecom approval was also granted to allow the fire protection equipment to be installed inside each cabinet of the PABX.

## TIMINGS

The following programme for the installation of the PABX was necessary to give British Telecom, the PABX supplier, contractors and site administration staff target dates:

| | |
|---|---:|
| Project approval | April 1981 |
| Obtain quotations and place orders | April/May 1981 |
| Builder's preparation work | September 1981 |
| Commence installation | October 1981 |
| Bring into service | 31st December 1981 |

## IMPLEMENTATION

The key factor in the implementation of our PABX was accommodation. Because of limited space within the office, it was proposed to install the new PABX in the same room as the existing PABX 7. Obviously the full co-operation of British Telecom was needed to achieve success.

The size of the PABX room is 17 feet by 13 feet maximum. Building work was negligible, but lighting and power outlets needed changing.

The sequence of events to install the PABX was as follows:

— Remove maintenance work bench to temporary room;

— Re-position PABX 7 batteries in temporary position;

— Remove PABX 7 doors and re-position cabinets in PABX room;

— Install SL1 peripheral equipment and MDF in permanent position;

— Locate SL1 common equipment in temporary position within PABX room;

— Bring SL1 into service;

— Disconnect PABX 7 and batteries and remove from room;

— Relocate common equipment and teletypewriter in permanent position;

— Install stand-by batteries;

— Re-site maintenance work table.

Total cost for reusing the same accommodation was less than £3,000. During the testing process, Reliance and British Telecom engineers worked together on a 'joint test' to reduce the commissioning/acceptance period.

## PROGRAMMING

In order to program the exchange for class of service. (Emergency, Telephones, Night Service, etc), data sheets have to be prepared. These include:

— PABX numbering scheme;

— Facility codes;

— Inter-PBX routes;

— System features;

— Call pick up groups;

— Hunting groups;

— Authorisation codes;

— Exchange lines;

— Emergency facilities;

— Short code dialling;

— Code restriction groups.

## IN SERVICE

The SL1 PABX came into service on schedule at the beginning of January 1982 and no major problems have occurred since then. A few minor difficulties have been experienced – Level 9 access (which was proven to be on the first 5 exchange lines in the BT main exchange) and no DDI on Night Service switching (caused by a fault on the DDI equipment).

Maintenance is provided by British Telecom: this is adequate, but it is obvious that they are still learning about the SL1 PABX. Maintenance charges for the PABX equipment are approximately £5,000 per annum. We have discovered that inputting and monitoring the information on the teletypewriter terminal is very time consuming, particularly during the early stages of PABX operation. Access onto our Company Internal Telephone Network was achieved when the PABX came into service, using E & M signalling interface cards. Electronic-type tones are generated by the SL1 but this difference, compared with the rest of our network, was not a significant change for the telephone users.

The implementation of the SL1 PABX on our network went smoothly and was trouble free, unlike the connection of some analogue PABXs where we experienced several problems, such as timing delays, tone suppression and inter-digit pause settings.

Overall, the site telephone users are very pleased with their new digital PABX and take advantage of the many facilities that it offers to them.

# 4 Using a Stored Program Control Exchange

M H Denny (Sea Containers Inc)

## INTRODUCTION

The planning prior to the installation of a stored program control voice switch needs to be every bit as detailed as the planning surrounding one of its more conventional predecessors. In addition to the normal circuit requirements, space and environment considerations, a wealth of detail surrounds the planning of the facilities. The use, explanation and allocation of the facilities at an early stage, including the involvement of the telephone supervisor, is necessary to lay the foundation for thorough planning, eventually leading to successful SPC exchange use.

## THE IMPACT OF AN SPC EXCHANGE ON THE ORGANISATION

The eventual successful use of the exchange by operators, supervisors and users alike will depend to a large extent upon the preparation prior to installation. Ideally, all users of the exchange will have been prepared for the variety of facilities during the pre-installation planning sessions. These will determine how the facilities will be used by each group of people.

As with any innovation, the SPC exchange will be strange to the users. Persuading them to use features which are unfamiliar is an on-going task. To ensure training continuity, user sessions aimed at people joining the company should be arranged periodically. An introduction to the company should cover the use of the telecommunications systems and enlightened personnel departments are only too delighted if the technical niceties of the telephone system

53

can be presented by the telecommunications manager or one of his staff. Once the features are understood and used, and the telephone and personnel people are playing their part in the presentation of the system to the users, the role of the telecommunications manager within the company begins to be understood.

Hand in hand with the training goes the documentation. The minimum of documentation required still necessitates considerable preparation. A descriptive booklet dealing with system use is a must. Another document which has to be prepared is a telephone directory covering everybody in the building or complex, together with details of any special routes to other switchboards which can be reached from the new system. A list of the abbreviated numbers and their prefixes must also be prepared. For the telephone supervisor, each extension should have a reference card listing the type of telephone, its location, department, user, class of service, and if it is part of a re-route or call pick-up group.

## THE IMPLICATIONS FOR THE TELECOMMUNICATIONS MANAGER

The telecommunications manager embarking on the installation of his first SPC exchange has a unique opportunity to re-position himself within the hierarchy of the company.

For the first time, the manager has the ability to overthrow the old image of cost centre management in favour of a fresh image of profit centre management. How is this achieved?

The process may be simplified by dividing it into two major actions: the maximisation of the facilities and the use of statistics. When facilities are maximised, the telephone switchboard operators become more involved in running the system because they are released from most of the chores of obtaining difficult calls, interconnecting extensions, and other peripheral activities. The difficult calls, ie long distance inter-continental calls, and other calls with only a limited success rate, will appear in the abbreviated dialling list which the user will be encouraged to maximise himself. Similarly, users will not be required to be connected to each other via the switchboard, because of the 'ring when free' facility. This facility will also cut down the incidence of unsuccessful calls. The 'Manager – Secretary' facility will eliminate the necessity of exchange Plan 107 systems and the consequential power requirements to drive them. The 'barring' and 're-route' facilities will re-direct calls into manned extensions thus minimising the operator time spent in attempting to connect calls to unmanned extensions. The resulting savings will show in the reduction in exchange lines and operator staffing levels.

The second of the two most valuable aids to the telecommunications manager is the availability of statistics from his machine. Leaving apart call information logging, the statistics available determine the exchange line loading which in itself is a valuable guide to allocating exchange lines to in-coming, both-way, or outgoing availability. Priority in most businesses is quite rightly given to the in-coming caller. However, the configuration and grouping of the exchange lines can be changed on the basis of the statistics (on traffic trends) revealed by the machine.

Other statistics available include the identification of the busy hour for particular types of call, and for the total system. The graphs (Figures 10 to 13) indicate the various peaks occurring through the day. The allocation of exchange lines to absorb the traffic generated is contained in tables so that for the grade of service required the number of circuits can be applied to the traffic offered to those circuits (Figures 14 and 15).

The onus is very much upon the telecommunications manager, once a computer controlled exchange is more than a planned event, to acquire the skills necessary to optimise equipment use and achieve savings.

Call information logging (whereby the statistics concerning the itemised usage from each extension can be listed under a variety of headings) is another valuable tool.

SPC exchanges record details of calls made, showing the extension from which the call is made, the time that the call originated, and the elapsed time and cost. From this information, a software package can be implemented which can present reports consolidating the information, and departments and groups of extensions can be charged for calls made. Costly calls, most normally called numbers, long duration calls and telephone abuse, can be identified. Departmental managers charged with the use/mis-use of 'phones

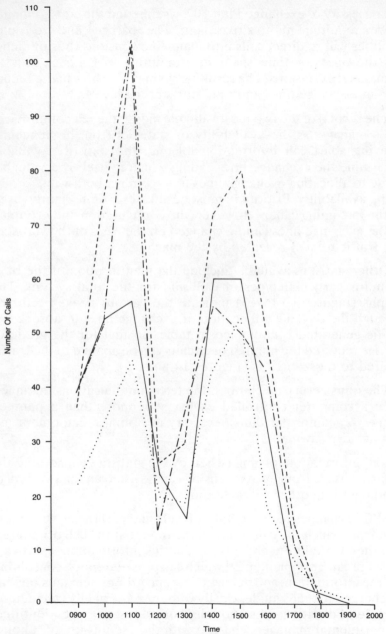

**Figure 10    Distribution of Calls Outgoing from London to Amsterdam**

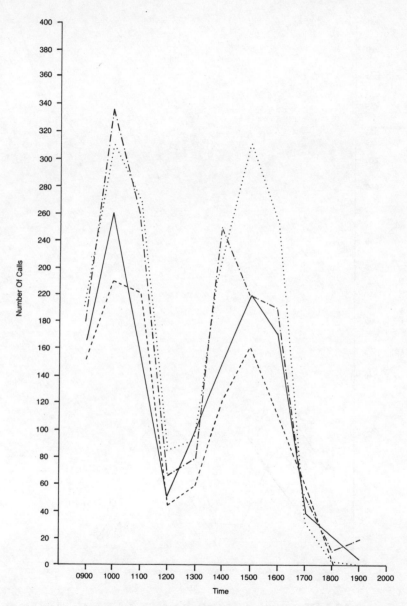

**Figure 11    Distribution of Calls Outgoing from London to Paris**

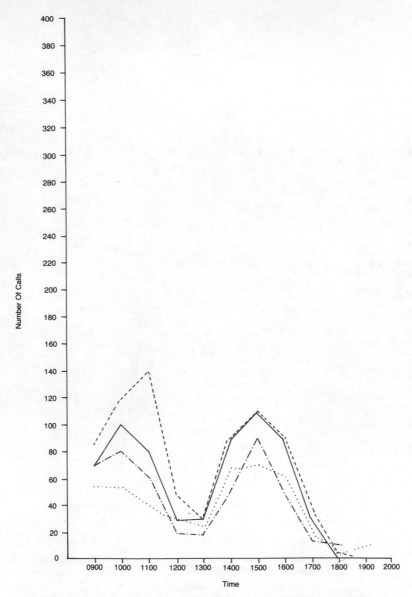

**Figure 12    Distribution of Calls Outgoing from London to Frankfurt**

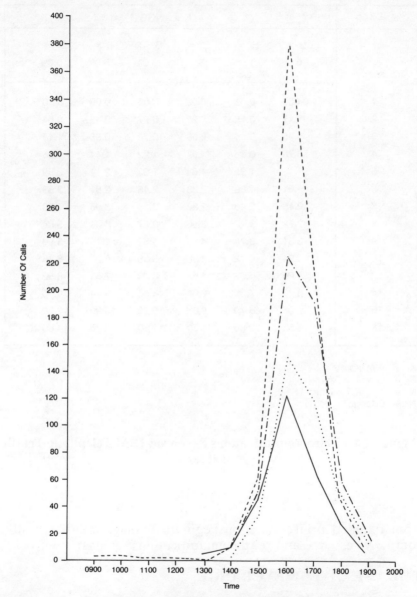

**Figure 13    Distribution of Calls Outgoing from London to San Francisco Monthly**

| K = 100 | A in ERLANGS | | | | | |
|---------|------------|------------|------------|------------|------------|------------|
| N | B = 0.5% | B = 1.0% | B = 2.0% | B = 3.0% | B = 5.0% | B = 10.0% |
| 1 | 0.005 | 0.01 | 0.02 | 0.03 | 0.05 | 0.10 |
| 2 | 0.11 | 0.15 | 0.22 | 0.27 | 0.36 | 0.54 |
| 3 | 0.35 | 0.46 | 0.59 | 0.70 | 0.86 | 1.14 |
| 4 | 0.70 | 0.86 | 1.07 | 1.22 | 1.44 | 1.85 |
| 5 | 1.12 | 1.35 | 1.63 | 1.82 | 2.11 | 2.59 |
| 6 | 1.61 | 1.89 | 2.23 | 2.46 | 2.81 | 3.58 |
| 7 | 2.15 | 2.48 | 2.88 | 3.15 | 3.55 | 4.20 |
| 8 | 2.72 | 3.10 | 3.56 | 3.87 | 4.31 | 5.04 |
| 9 | 3.31 | 3.74 | 4.25 | 4.61 | 5.10 | 5.90 |
| 10 | 3.94 | 4.42 | 4.98 | 5.36 | 5.91 | 6.76 |
| 12 | 5.25 | 5.82 | 6.48 | 6.93 | 7.55 | 8.52 |
| 14 | 6.63 | 7.28 | 8.04 | 8.54 | 9.24 | 10.30 |
| 16 | 8.06 | 8.79 | 9.63 | 10.20 | 10.90 | 12.20 |
| 18 | 9.53 | 10.30 | 11.30 | 11.80 | 12.70 | 14.00 |

K = Availability
B = Blocking Rate
N = No. of Tie Lines
A = Erlangs

**Figure 14   Telecommunications Research IBM Telephone Traffic Tables**

from detailed print-outs soon begin to manage another costly company resource and reap communications savings.

## THE USE OF THE EXCHANGE

The installed exchange, in most cases, is used solely as a voice switchboard, while data transmission continues on direct exchange lines. Why not integrate the two? The wiring for which BT charges

| K = 10 | A IN ERLANGS | | | | | | | | |
|---|---|---|---|---|---|---|---|---|---|
| N | B = 0.1% | B = 0.2% | B = 0.5% | B = 1% | B = 2% | B = 3% | B = 5% | B = 10% | B = 20% |
| 1 | 0.00 | 0.00 | 0.00 | 0.01 | 0.02 | 0.03 | 0.05 | 0.11 | 0.25 |
| 2 | 0.05 | 0.07 | 0.11 | 0.15 | 0.22 | 0.28 | 0.38 | 0.60 | 1.00 |
| 3 | 0.19 | 0.25 | 0.35 | 0.46 | 0.60 | 0.72 | 0.90 | 1.27 | 1.93 |
| 4 | 0.44 | 0.54 | 0.70 | 0.87 | 1.09 | 1.26 | 1.52 | 2.05 | 2.94 |
| 5 | 0.78 | 0.90 | 1.13 | 1.36 | 1.66 | 1.88 | 2.22 | 2.88 | 4.01 |
| 6 | 1.06 | 1.33 | 1.62 | 1.91 | 2.28 | 2.54 | 2.96 | 3.76 | 5.11 |
| 7 | 1.08 | 1.80 | 2.16 | 2.50 | 2.94 | 3.25 | 3.74 | 4.67 | 6.23 |
| 8 | 2.05 | 2.31 | 2.73 | 3.13 | 3.63 | 3.99 | 4.54 | 5.60 | 7.37 |
| 9 | 2.66 | 2.86 | 3.33 | 3.78 | 4.34 | 4.75 | 5.37 | 6.55 | 8.52 |
| 10 | 2.09 | 3.43 | 3.96 | 4.46 | 5.08 | 5.53 | 6.22 | 7.51 | 9.69 |
| 11 | 3.65 | 4.02 | 4.61 | 5.16 | 5.84 | 6.33 | 7.08 | 8.49 | 10.90 |
| 12 | 4.23 | 4.64 | 5.28 | 5.88 | 6.61 | 7.14 | 7.95 | 9.47 | 12.00 |
| 13 | 4.83 | 5.27 | 5.96 | 6.61 | 7.40 | 7.97 | 8.83 | 10.50 | 13.20 |
| 14 | 5.45 | 5.92 | 6.66 | 7.35 | 8.20 | 8.80 | 9.73 | 11.50 | 14.40 |
| 15 | 6.08 | 6.58 | 7.38 | 8.11 | 9.01 | 9.65 | 10.60 | 12.50 | 15.60 |
| 16 | 6.78 | 7.26 | 8.10 | 8.88 | 9.83 | 10.50 | 11.50 | 13.50 | 16.80 |
| 17 | 7.38 | 7.95 | 8.83 | 9.65 | 10.70 | 11.40 | 12.50 | 14.50 | 18.00 |
| 18 | 8.05 | 8.64 | 9.58 | 10.40 | 11.50 | 12.20 | 13.40 | 15.50 | 19.20 |
| 19 | 8.72 | 9.35 | 10.30 | 11.20 | 12.30 | 13.10 | 14.30 | 16.60 | 20.40 |
| 20 | 9.41 | 10.10 | 11.10 | 12.00 | 13.20 | 14.00 | 15.20 | 17.60 | 21.60 |

K = Availability
B = Blocking Rate
N = No. of Tie Lines
A = Erlangs

**Figure 15    Telecommunications Research ITT Telephone Traffic Tables**

handsomely is provided throughout the building, and it is a local area network through which all things can be connected.

The final phase of this chapter deals with some of the less obvious aspects of using any telephone exchange, particularly the computer controlled type.

Environmental considerations determine where the equipment is to be situated. The larger machines will need a false floor computer style of environment with air conditioning. Floor loading and the route into the building that the equipment must take upon delivery must be considered. Floor loadings along the route, weight of equipment and maximum permissible weight limits of any lifts to be used, must all be known in advance to prevent exasperating situations arising. Stand-by power, back up, and the provision of batteries should also be considered at the planning stage, along with the location of the telephone operator, if one is necessary.

The power supply and questions concerning the results of power failure, together with the need for stand-by generators or batteries, should be considered early in the planning. Maintenance contracts, the documentation and the on-going technical responsibility need to be understood, and guidelines and procedures drafted.

When the system is fully operational, the expansion potential and cost of enhancing and integrating the various systems need to be fully understood, and the cost information continually monitored. In this context, the successful telecoms manager will have installed his system and be charging the users upon the basis of call information logging, so that not only is the system financially self-supporting, but it also pays the salaries and overheads of the telecommunications staff. The savings pruned from the operational margins will initially be used to finance the integration and enhancing of the system. In short, the beginnings of the transition from cost centre to profit centre.

# Appendix 1

## PABX Models and their Capacities
### (April 1982)
J E Lane
(NCC Office and Communications Systems Division)

# PABX MODELS

PABX >100 extns

| Supplier | Model | Type | Capacity (trunks/extns) |
|----------|-------|------|-------------------------|
| BT | Regent | SPC electronic | 24/134 |
| | Monarch | SPC digital | 30/120 |
| Ferranti-GTE | GTD-1000E * | SPC digital | 256/1024 |
| | GTD-4600E * | SPC digital | 4600/12200 |
| Harris | D1201 * | SPC digital | 520 † |
| | D1202 * | SPC digital | 1024 † |
| | D1203 * | SPC digital | 144 † |
| IBM | 1750 | SPC electronic | 96/760 |
| | 3750 | SPC electronic | 356/2500 |
| ICL | DNX2000 * | SPC digital | 10000 † |
| ITT | Pentomat 200 | Crossbar | 44/200 |
| | Pentomat 1000CT | Crossbar | 109/1000 |
| | Pentomat 1000T2 | Crossbar | as req/9000 |
| | 4080 Unimat | SPC electronic | 10000 † |
| Mitel/Norton | SX2000 * | SPC digital | 10000 † |
| Plessey | PDX | SPC digital | 700/4000 |
| | ACD | SPC digital | 60 to 120 agents |
| Pye/TMC (Philips) | UH 200 | Rotary switch | 35/200 |
| | UH 900 | Rotary switch | 60/800 |
| | EBX 8000 | SPC reed relay | 900/8000 |
| Reliance (GEC) | SL-1 | SPC digital | 950/5000 |
| Telephone Rentals | PDX | SPC digital | 120/800 |
| Thorn-Ericsson | ARD 561 | Crossbar | 40/270 |
| | ARD 562 | Crossbar | 100/540 |
| | AKD 791/2/3 | Code switch | 450/9000 |
| | MD 110 * | SPC digital | 10000 † |

† total including extensions and exchange lines
* awaiting BT approval

# Appendix 2

## PABX Suppliers
J E Lane (NCC)

British Telecom
    Local Telephone Sales Office
    (address and telephone number in preface to telephone direc-
    tory)

Ferranti GTE Ltd.,
    Head Office:
    St. Mary's Road,
    Moston,
    Manchester M10 0BE                      Tel: 061 681 2071

    London Office:
    110 Euston Road,
    London NW1 2DQ                          Tel: 01 387 9770

Harris Systems Ltd.,
    153 Farnham Road,
    Slough,
    Berks SL1 4XD                           Tel: 0753 34666

IBM UK Ltd.,
    IBM Telephone Switching Systems,
    389 Chiswick High Road,
    London W4 4AL                           Tel: 01 995 1441

ICL,
    Computer House,
    292/298 High Street,
    Slough SL1 4NA                          Tel: 0753 3111

ITT Business Systems,
    Lion Buildings,
    Crowhurst Road,
    Hollingbury,
    Brighton,
    Sussex BN1 8AN                          Tel: 0273 507111

Mitel Telecom Ltd.,
    33/37 Queen Street,
    Maidenhead,
    Berks SL6 1NB                           Tel: 0628 72821

Norton Telecommunications,
    Norton House,
    Bilton Way,
    Luton                      Tel: Luton 416444

Philips Business Systems,
    Communications and Control Division,
    Cromwell Road,
    Cambridge CB1 3HE          Tel: 0223 245191

Plessey Communication Systems,
    Beeston,
    Nottingham NG9 1LA        Tel: 0602 254822

Reliance Systems Ltd.,
    Turnells Mill Lane,
    Wellingborough,
    Northants. NN8 2RB        Tel: 0933 225000

Telephone Rentals,
    TR House,
    Bletchley,
    Milton Keynes MK3 5JL     Tel: 0908 71200

Thorn-Ericsson,
    Telecommunications (Sales) Ltd.,
    Viking House,
    Foundry Lane,
    Horsham,
    West Sussex RH13 5QF     Tel: 0403 64166

# Appendix 3

## PABX Facilities
J E Lane (NCC)

The following list describes features and services commonly found on most SPC PABXs.

Sometimes the same facility is given different names by different suppliers; sometimes similar names conceal minor differences in a facility or service.

The list does not purport to be definitive or comprehensive.

| | |
|---|---|
| Abbreviated Dialling | Enables calls to be placed to outside subscribers by dialling a code (2 or 3 digits). The PABX provides a translation of digits to be pulsed out to the public network. |
| Absent Transfer | Automatically routes incoming calls to a pre-selected extension when the called directory number does not answer within a prescribed time. |
| Alternative Trunk Routeing | Automatically routes outgoing calls over alternative facilities when all trunks in a trunk group are busy. |
| Automatic Recall | Operator receives an automatic recall signal on any incoming call still uncompleted after a given period of time (see also Camp-on-Busy). |
| Barring | Synonymous with Call Restriction. |
| Break-in | Allows the operator to break into an existing extension conversation; a warning tone is first transmitted to the extension to indicate the pending override. |
| Broker's Enquiry | Allows an extension user on an established two-party call to hold the existing call and originate another call to a third party for private consultation. The extension user can then alternate between the two parties without further dialling. |

| | |
|---|---|
| Call Announcing | An incoming call may be held while the operator announces it to the extension (see also Call Splitting). |
| Callback | Enables an extension user to overcome the problem of having to repeatedly re-dial an extension that is engaged. The user selects a code requesting the facility and the system automatically rings both parties when the called party becomes free. Callback may also be provided for outgoing trunk calls and trunk services. |
| Call Distribution Group | Allows an incoming call to be routed automatically without operator assistance to a group of assigned extensions on a rotational basis (see also Hunting). |
| Call Diversion | Synonymous with Call Forward. |
| Call Enquiry | see Call Hold. |
| Call Forward | Allows calls routed to a specific extension to be rerouted automatically to another extension or the operator. Call Forward (Follow Me) will route all calls regardless of the state of the original extension; Call Forward (Busy/No Answer) will route calls only when the station is busy or does not answer within a prescribed period of time. |
| Call Hold | Allows an extension user to place an existing call in a 'hold' mode, make an enquiry call to another extension or operator, and then return to the held call. The call on hold is isolated from the extension and cannot join in conversations unless a Conference (add – on) is established (see also Broker's Enquiry). |
| Call Offering | Synonymous with Trunk Offering. |

Call Pick-Up

Allows an extension user to answer calls directed to another extension. Directed Call Pick-Up allows an extension to answer by dialling a Pick-Up code plus the extension's number. Group Call Pick-Up allows the extension to answer calls to other extensions in the same Call Pick-Up Group by dialling the Pick-Up code only.

Call Restriction

Denies or allows selected extension and tie trunk users access to international, national and local public exchange codes. Call restriction facilities include: unrestricted; trunk restricted; semi-restricted; and fully restricted.

Call Splitting

Enables the operator to split the calling and called parties and speak privately to either (see also Call Announcing).

Call Transfer

Allows an extension user to transfer an existing call to another extension or the operator. A transfer is usually effected from a Call Hold condition when the original extension dials the other extension and then hangs up.

Camp-on Busy

Enables a call to a busy extension to be held waiting by the operator (camped-on) until the called line becomes idle. When the line becomes idle the called extension is rung automatically. If the camp-on call is not answered within a set period of time, the Automatic Recall alerts the operator.

Class of Service

Synonymous with Classification of Extensions.

| | |
|---|---|
| Classification of Extensions | Different combinations of classes of service to control calls originating from extensions or Tie-lines to the public network and special services (see also Call Restriction). |
| Conference (Add-on) | Allows an extension user to establish a three-way conference with a held call and another station. Some systems allow an Add-on link to another outgoing trunk as well as the internal extension. |
| Conference (Operator controlled) | Allows the operator to establish a conference between a public exchange line or Inter-PBX line and a pre-defined number of extensions. |
| Dictation Access | Allows extension users access to and control of centralised dictation equipment. |
| Direct Inward Dialling (DID) | Allows an incoming call from the exchange network to reach a specific extension without assistance from the operator. |
| Discriminatory Ringing | Exchange lines and internal calls have different ringing cycles. |
| Do-not-Disturb | Allows the user to place an extension into a mode in which extensions may originate calls but appear busy to all incoming calls. |
| Emergency ('meet-me') Conference | Allows a number of extensions to take part in a conference by dialling a special code. |
| Emergency Transfer | Allows two-way service to exchange network for a limited number of pre-arranged extensions during a system or power failure. |
| Enquiry Call | Synonymous with Call Hold. |

| | |
|---|---|
| Executive Intrusion | Allows an extension user to cut into the conversation of an engaged extension by dialling a code. Warning tone advises the conversing parties that their conversation is being intruded upon. |
| Flexible Numbering | Directory numbers may be assigned to extensions during installation in accordance with users' numbering plan. |
| Follow-me | Synonymous with Call Forward. |
| Group Calling/ Hunting | Synonymous with Hunting – terminal-circular. |
| Group Call Pick-up | See Call Pick-up. |
| Hold for Enquiry | Synonymous with Call Hold. |
| Holding | Synonymous with Parking of Calls. |
| Hot-line | Enables specified extensions to be immediately connected to a predetermined extension or external subscriber. |
| Hotel Features | Hotel Features are optional on many systems. These include: room status; message waiting; and automatic wake-up. |
| Hunting | Routes an incoming call to an idle extension in a pre-arranged group, when the called extension is busy. Hunting facilities include: terminal; circular and secretarial. |
| Inquiry | Synonymous with Call Hold. |
| Inter-PABX Connect | Equipment provided to connect the system to a distant PABX via private circuits (see also Tie-lines). |
| Least Cost Routeing | Automatically routes outward calls over trunk facilities on an optimised least cost basis. |

| | |
|---|---|
| Line Lockout | A handset accidentally left off-hook is automatically switched to a lockout position after a predetermined length of time. |
| Night Service | The Night Service facility allows for unattended operation of the PABX at night, or during weekends, holidays, etc. |
| Operator Intrusion | Synonymous with Break-in. |
| Operator Recall | Enables an extension engaged on an exchange or Inter-PBX call to signal the operator for assistance. |
| Paging | Paging facilities allow the operator or extension access to loudspeaker public address or radio paging systems. |
| Parking of Calls | Enables the operator to park an incoming call while attending to other calls. After a specific period of time, Automatic Recall alerts the operator to attend to the call. |
| Priority | Synonymous with Executive Intrusion. |
| Push-button Dialling | Allows the use of telephones equipped with push-button dials to transmit digits via audible tones to the switching equipment. |
| Repeated Enquiry | Synonymous with Broker's Enquiry. |
| Route Advance | Synonymous with Alternative Trunk Routeing. |
| Secretarial Service | Allows any two extensions to be associated in a manager/secretary arrangement (see also Hunting – secretarial). |
| Serial Calls | Where an outside caller wishes to speak to several different extensions, this feature enables the caller to be reconnected automatically to the operator after each call for transfer to the next extension. |

| | |
|---|---|
| Speed Dialling | Synonymous with Abbreviated Dialling. |
| Tie-lines | Provide interconnections between several private exchanges (see also Inter-PABX Connect). |
| Trunk Offering | Enables the operator to interrupt an existing conversation to offer an important new call. A warning tone will be heard by the conversing parties before interruption. |

# Appendix 4

## SPC PABX Features Comparative List
J E Lane (NCC)

| SUPPLIER<br>MODEL | British Telecom<br>REGENT | British Telecom<br>MONARCH | Ferranti GTE<br>GTD-1000E | Ferranti GTE<br>GTD-4600E |
|---|---|---|---|---|
| *SWITCHING*<br>*TECHNOLOGY* | analogue<br>space division<br>electronic | digital<br>time division<br>electronic<br>PCM | digital<br>time division<br>electronic<br>PCM | digital<br>time division<br>electronic<br>PCM |
| *SYSTEM CAPACITY*<br>Extensions<br>Exchange lines (trunks)<br>Inter-PBX tie lines | 134<br>24<br>2 | 20-120<br>} 34 | 1024<br>}256 | 12288<br>}4608 |
| Traffic capacity<br><br>Traffic density<br>per extn. (erlangs) | 31 simultaneous<br>calls<br><br>– | 32 simultaneous<br>calls<br><br>0.22 | 192 simultaneous<br>calls<br><br>0.23 | 2304 simultaneous<br>calls<br><br>0.25 |
| *COST per extn.*<br>(depends on<br>configuration) | lease/rental<br>only | rental or lease<br>only | £425-500 | £300-500 |
| *FEATURES*<br>Dial phones<br>MF keyphones<br>Data terminals/max. speed | Yes<br>Yes<br>Yes 9.6 kbit/s | Yes<br>Yes<br>Yes 9.6 kbit/s | Yes<br>Yes<br>Yes 64 kbit/s | Yes<br>Yes<br>Yes 64 kbit/s |
| Simultaneous<br>  voice/data on same<br>  extension circuit | Yes | future | Yes | Yes |
| Exchange line<br>  signalling:<br>– dial 10pps;<br>– multi frequency (MF4);<br>– PCM 2Mbit/s | Yes<br>future<br>future | Yes<br>future<br>future | Yes<br>Yes<br>Yes | Yes<br>Yes<br>Yes |
| Tie Line signalling:<br>– loop disconnect;<br>– SCDC;<br>– AC13;<br>– AC15;<br>– E & M;<br>– MF5 inter-register. | Yes<br>Yes<br>Yes<br>Yes<br>Yes<br>No | No<br>Yes<br>Yes<br>Yes<br>Yes<br>No | Yes<br>Yes<br>No<br>Yes<br>Yes<br>No (alternative<br>available) | Yes<br>Yes<br>No<br>Yes<br>Yes<br>No (alternative<br>available) |
| Main/satellite working | Yes | Yes | Yes | Yes |
| Control system:<br>Processor | single micro | single micro | duplicate<br>(optional) | triplicate |
| Memory storage medium<br>– main;<br>– back-up | RAM/ROM<br>battery powered<br>RAM | RAM/PROM<br>battery powered<br>RAM | RAM/ROM<br>mag. tape | RAM/ROM<br>mag. tape |

| SUPPLIER<br>MODEL | Harris<br>D1200 Series | | | IBM<br>3750 | IBM<br>1750 |
|---|---|---|---|---|---|
| *SWITCHING*<br>*TECHNOLOGY* | Digital<br>time divison electronic<br>delta modulation | | | Analogue<br>space division<br>electronic | Analogue<br>space division<br>electronic |
| *SYSTEM CAPACITY*<br>Extensions<br>Exchange lines (trunks)<br>Inter-PBX tie lines | *D1203*<br><br>} 144 | *D1201*<br><br>520 | *D1202*<br>2500<br>1024 }228 | 760<br><br>}96 | |
| Traffic capacity | 140 simultaneous calls | | | 378 simult. calls | 144 simult. calls |
| Traffic density<br>per extn. (erlangs) | non-blocking 0.5 | | 0.2 | to suit require-<br>ments | to suit require-<br>ments |
| *COST per extn.*<br>(depends on<br>configuration) | £400 - 600 | | | £400 - 700 | £400 - 700 |
| *FEATURES*<br>Dial phones<br>MF keyphones<br>Data terminals/max. speed<br>Simultaneous<br>  voice/data on same<br>  extension circuit<br>Exchange line<br>  signalling:<br>–  dial 10pps;<br>–  multi frequency (MF4);<br>–  PCM 2Mbit/s. | Yes<br>Yes<br>Yes 9600 bit/s<br><br>early 1983<br><br><br><br>Yes<br>Yes<br>future | | | Yes<br>Yes<br>Yes<br><br>No<br><br><br><br>Yes<br>No<br>No | Yes<br>Yes<br>Yes<br><br>No<br><br><br><br>Yes<br>No<br>No |
| Tie line signalling:<br>–  loop disconnect;<br>–  SCDC;<br>–  AC13;<br>–  AC15;<br>–  E & M;<br>–  MF5 inter-register. | Yes<br>Yes<br>No<br>Yes<br>Yes<br>field trials in progress | | | Yes<br>Yes<br>Yes<br>No<br>Yes<br>Yes | Yes<br>Yes<br>Yes<br>Yes<br>Yes<br>Yes |
| Main/satellite working | Yes | | | Yes | Yes |
| Control system:<br>Processor | duplicate (optional) | | | duplicate | duplicate |
| Memory storage medium<br>–  main;<br>–  back-up | RAM, PROM<br>tape cartridge | | | FET<br>mag. disk | FET<br>mag. disk |

| SUPPLIER<br>MODEL | ITT<br>Unimat 4080 | Mitel/Norton<br>SX2000<br>ICL DHX 2000 | Philips<br>EBX 8000 | Plessey/<br>Telephone<br>Rentals PDX |
|---|---|---|---|---|
| *SWITCHING TECHNOLOGY* | analogue<br>space division<br>electronic | space division<br>electronic<br>PCM | digital<br>space division<br>reed relays | time division<br>electronic<br>PCM |
| *SYSTEM CAPACITY*<br>Extensions<br>Exchange lines (trunks)<br>Inter-PBX tie lines | 150-8000<br>}704 | }150-10000 | 8000<br>}900 | 4000<br>} 700 |
| Traffic capacity | as required | — | — | 450 sim. calls<br>minumum |
| Traffic density<br>per extn. (erlangs) | 0.18 | 0.3 (analogue)<br>0.6 (digital) | 0.24 | 0.18 |
| *COST per extn.*<br>(depends on<br>configuration) | £350 - 400 | £300 - 350 | £350 - 400 | £300 - 700 |
| *FEATURES*<br>Dial phones<br>MF keyphones<br>Data terminals/max. speed<br>Simultaneous<br>  voice/data on same<br>  extension circuit<br>Exchange line<br>  signalling:<br>–  dial 10pps;<br>–  multi frequency (MF4);<br>–  PCM 2Mbit/s. | Yes<br>Yes<br>Yes 19.2 Kbit/s<br><br>Yes<br><br><br>Yes<br>Yes<br>Yes | Yes<br>Yes<br>Yes<br><br>Yes<br><br><br>Yes<br>Yes<br>Yes | Yes<br>Yes<br>Yes<br><br>No<br><br><br>Yes<br>Yes<br>Yes | Yes<br>Yes<br>Yes 64 Kbit/s<br><br>Yes<br><br><br>Yes<br>Yes<br>Yes (in future) |
| Tie line signalling:<br>–  loop disconnect;<br>–  SCDC;<br>–  AC13;<br>–  AC15;<br>–  E & M;<br>–  MF5 inter-register. | Yes<br>Yes<br>Yes<br>Yes<br>Yes<br>No | Yes<br>Yes<br>Yes<br>Yes<br>Yes<br>Yes | Yes<br>Yes<br>Yes<br>Yes<br>Yes<br>Yes | Yes<br>Yes<br>Yes<br>Yes<br>Yes<br>Yes |
| Main/satellite working | No | Yes | No | Yes |
| Control system:<br>Processor | distributed | distributed<br>micros | duplicate | duplicate |
| Memory storage medium<br>–  main;<br><br>–  back-up | RAM/PROM<br><br>mag. disk | bubble memory/<br>RAM<br>not required | RAM<br><br>paper tape | MOS<br><br>mag. tape<br>battery<br>powered RAM |

| SUPPLIER<br>MODEL | Reliance (GEC)<br>SL1 | Thorn Ericsson<br>MD110 |
|---|---|---|
| *SWITCHING*<br>*TECHNOLOGY* | digital<br>TDM<br>electronic<br>PCM | digital (time<br>division)<br>electronic<br>PCM |
| *SYSTEM CAPACITY*<br>Extensions<br>Exchange lines (trunks)<br>Inter-PBX tie lines | 5000+<br>} 952 | } 10000+ |
| Traffic capacity<br><br>Traffic density<br>per extn. (erlangs) | 1020 simult. calls<br><br><br>variable | 2000 busy hour<br>calls per module<br><br>0.25E/extn. |
| *COST per extn.*<br>(depends on<br>configuraton) | £360 - 1000 | £350 - 700 |
| *FEATURES*<br>Dial phones<br>MF keyphones<br>Data terminals/max. speed<br>Simultaneous<br>  voice/data on same<br>  extension circuit<br>Exchange line<br>  signalling:<br>–  dial 10pps;<br>–  multi frequency (MF4);<br>–  PCM 2Mbit/s<br><br>Tie line signalling:<br>–  loop disconnect;<br>–  SCDC;<br>–  AC13;<br>–  AC15;<br>–  E & M;<br>–  MF5 inter-register.<br><br>Main/satellite working<br><br>Control system:<br>Processor<br><br>Memory storage medium<br>–  main;<br>–  back-up | Yes<br>Yes<br>Yes 19.2 Kbit/s<br><br>Yes<br><br><br><br>Yes<br>future<br>future<br><br><br>Yes<br>Yes<br>Yes<br>Yes<br>Yes<br>Yes<br><br>Yes<br><br><br>duplicate<br><br><br>RAM/ROM<br>mag. tape | Yes<br>Yes<br>Yes<br><br>Yes<br><br><br><br>Yes<br>future<br>future<br><br><br>Yes<br>Yes<br>Yes<br>Yes<br>Yes<br>future<br><br>Yes<br><br><br>distributed<br><br><br>RAM/EPROM<br>mag. tape cartridge |

| SUPPLIER MODEL | British Telecom REGENT | British Telecom MONARCH | Ferranti GTE GTD-1000E | Ferranti GTE GTD-4600E |
|---|---|---|---|---|
| **FACILITIES** | | | | |
| User facilities: standard/optional | package or tailored individually | standard and additional opt. | all standard | all standard |
| Changing of facilities | supplier, via console | by user, via console | user | user |
| System logging Call logging | Yes | console display | Yes | Yes |
| – interface for external equipment | Yes | Yes | Yes | Yes |
| **INSTALLATION** | | | | |
| Extns./trunks per cabinet | 134/24 | 120/34 | 256/64 | — |
| Extn./trunk module size | 8/4 | 4/2 | 8/4 | — |
| Extn. limits | 1000 ohm | 1200 ohm | 2000 ohm | 2000 ohm |
| Console separation | 300 metres | 300 metres + | unlimited | unlimited |
| Console | desk-top | desk-top | desk-top | desk-top |
| Cabinet dimensions (d x w x h) | 700x600x986mm | 0.6x0.6x1.7 metre | 28in x 25in x72in | 27in x22in x 96in |
| Cabinet weight | 132 kg | — | 274 kg | — |
| Power supply | 240v a.c. | 240v 50Hz | 50v d.c. | 50v d.c. |
| Max. consumption | 500w | 600w | 3kw | — |
| Cabinet dissipation | 500w | 600w max. | — | — |
| Temp. range | office environment | — | office environment | — |
| **GENERAL** | | | | |
| BT approval | approved | approved | awaiting | awaiting |
| Delivery | 5-8 weeks | 3 weeks | 4 months | 6-9 months |
| Installation | depends on wiring | depends on wiring | 4-6 weeks | — |
| BT acceptance | n/a | n/a | 4 weeks | — |
| Number installed – in UK – eleswhere | — — | — — | 0 (approx. 10000 total) | 1 |
| Country of origin | UK | UK | US | US |
| Country of manufacture | UK | UK | US/Belgium/UK (after 1982) | US/UK |
| Documentation and training | consult supplier | consult supplier | consult supplier | consult supplier |

| SUPPLIER<br>MODEL | Harris<br>D1200 Series | | | IBM<br>3750 | IBM<br>1750 |
|---|---|---|---|---|---|
| **FACILITIES**<br>User facilities:<br>standard/optional | Extn. facilities all standard<br>Networking facilities<br>optional | | | individually<br>tailored | package options |
| Changing of facilities | User, via terminal/switch<br>settings | | | by user, via<br>terminal | by user, via<br>terminal |
| System logging<br>Call logging<br>– interface for<br>external equipment | mag. tape<br><br>Yes | | | mag. disk<br><br>Yes (bi-synch) | mag. disk<br><br>Yes (bi-synch) |
| **INSTALLATION**<br>Extns./trunks per<br>cabinet | *D1203*<br><br>144 | *D1201*<br><br>520 | *D1202*<br><br>1024 | — | — |
| Extn./trunk module size | 8 | 8 | 8 | 4/2 | 4/2 |
| Extn. limits | 600 ohm/1200 ohm | | | 760, 1200 ohm | 760, 1200 ohm |
| Console separation | 1000 ft. | | | 152m | 350m |
| Console | desk-top | | | stand-alone | desk-top |
| Cabinet dimensions<br>(d x w x h) | 27x24x36in | 27x24x71in | 27x46x71in | 3x (1750 size) | 4mx84cmx<br>178cm |
| Cabinet weight | 300 1b | 650 1b | 1250 1b | 3x(1750 wt) | 1950 kg |
| Power supply | 240v 50Hz or -48 volt | | | 240v 50Hz<br>only | 240v 50Hz,<br>50v |
| Max. consumption | 1kVA | 2kVA | 4kVA | 12-50 kw | 3 kw |
| Cabinet dissipation | — | | | 15kVA | 4kw |
| Temp. range | 0-55°c | | | 15-38°c | 10-46°c |
| **GENERAL**<br>BT approval | expected end 1982 | | | approved | approved |
| Delivery | 60 days | | | 5-6 months | 5-6 months |
| Installation | 1 week (200 lines) | | | 2-3 weeks | 2 weeks |
| BT acceptance | 1-2 weeks | | | 6 weeks | 6 weeks |
| Number installed<br>– in UK<br>– elsewhere | field trials only<br>6000+ | | | (over 300 total)<br>— | |
| Country of origin | US | | | France | France |
| Country of manufacture | US/UK/Europe | | | France | France |
| Documentation and<br>training | consult<br>supplier | | | consult<br>supplier | consult<br>supplier |

# PABX COMPARATIVE LIST

| SUPPLIER MODEL | ITT Unimat 4080 | Mitel/Norton SX2000 ICL DHX 2000 | Philips EBX 8000 | Plessey/ Telephone Rentals PDX |
|---|---|---|---|---|
| **FACILITIES** User facilities: standard/optional | all standard | all standard | all standard | package opts. |
| Changing of facilities | by user, via terminal | by user, via terminal | user | user |
| System logging | mag. disk/tape | mag. disk | paper tape or print-out | Yes |
| Call logging – interface for external equipment | Yes | Yes | Yes | Yes |
| **INSTALLATION** Extns./trunks per cabinet | 200/24 | 768 ports (first cabinet) | 250/32 | 200-300 |
| Extn./trunk module size | 8/2 | 16/8 | 128/16 | 8/4 |
| Extn. limits | 1000 ohm (without tel) | 1000 ohm | 1200 ohm | 1200 ohm |
| Console separation | 600 metres | 5000 feet | 300 m | 300 m |
| Console | stand-alone | desk-top | desk-top | desk-top |
| Cabinet dimensions (d x w x h) | 454x1000x 1800 mm | 720x816x 1656 mm | 460x900x 2200 mm | 650x1340x 1730 mm |
| Cabinet weight | 300 kg/m² | 273 kg | 300 kg | 520 kg |
| Power supply | 240v 50Hz/ 48v d.c. | 230v 50Hz -50v d.c. | 240v a.c./ 50v d.c. | 240v a.c./ 50v d.c. |
| Max. consumption | 6w per line | 1.2 kw max. | 3 kw | 2.5 kw |
| Cabinet dissipation | 0.75 kw | 700w | 200w | 1.3 kw |
| Temp. range | 5-35°c | 0-50°c | -5°c-45°c | 0°c-45°c |
| **GENERAL** BT approval | approved | expected early 1983 | approved | approved |
| Delivery | 20 weeks | 2 weeks | 20-30 weeks | immediate |
| Installation | 6-12 weeks | 6 weeks | 6-12 weeks | 4 weeks |
| BT acceptance | 6 weeks | 6 weeks | 4 weeks | 1 week |
| Number installed – in UK – elsewhere | 170 6000 | field trials — | 80 — | 300 6000+ |
| Country of origin | Europe/UK | UK/Canada | Holland | US |
| Country of manufacture | hardware – Germany software - UK | Canada. UK (after 1985) | Holland/UK | UK |
| Documentation and training | consult supplier | consult supplier | consult supplier | consult supplier |

| SUPPLIER<br>MODEL | Reliance (GEC)<br>SL1 | Thorn Ericsson<br>MD110 |
|---|---|---|
| *FACILITIES*<br>User Facilities:<br>standard/optional | all available<br>or denied every<br>ext | all standard |
| Changing of facilities | user | by user, via<br>terminal |
| System logging | print-out | print-out |
| Call logging<br>– interface for<br>external equipment | Yes | Yes |
| *INSTALLATION*<br>Extns./trunks per cabinet | variable | 252/84 |
| Extn./trunk module size | 4/2 | 6/2 |
| Extn. limits | 1000 ohm<br>(without tel) | 1800 ohm |
| Console separation | 4000 feet | 1000 metres |
| Console | desk-top | desk-top |
| Cabinet dimensions<br>(d x w x h) | 510x1320x1810mm | 300x600x1800mm |
| Cabinet weight | 300 kg | 200 kg |
| Power supply | 240v ac/50v dc | 240v 50Hz, 48v dc |
| Max. consumption | 4.8 kVA/1000 lines<br>average | 450w |
| Cabinet dissipation | 4.5 Btu/hr | 400w |
| Temp. range | 0-50°c | office environment |
| *GENERAL*<br>BT approval | approved | awaiting |
| Delivery | 16-20 weeks | 14 months |
| Installation | 6-8 weeks | 3-12 weeks |
| BT acceptance | 4-6 weeks | 6-12 weeks |
| Number installed<br>– in UK<br>– elsewhere | 100+<br>3000+ | field trials<br>— |
| Country of origin | Canada/UK | Sweden/UK |
| Country of manufacture | UK | Sweden (UK in 1983) |
| Documentation and<br>training | consult<br>supplier | consult<br>supplier |

## CHECKLIST NOTES

These notes provide a fuller explanation of the information contained in the list of SPC PABX features, in connection with system capacity, program storage and power supplies.

### 1.  SYSTEM CAPACITY

**Traffic Capacity**

Is an indication of system performance expressed as:-
   Number of simultaneous calls, ie the *maximum* number of conversations which can be handled by the PABX at a given time.

**Traffic Density per Extension (Erlangs)**

Is a measurement of the average occupancy of the extension terminal, where occupancy is the length of time the terminal is busy. One call lasting one hour corresponds to 1 erlang of traffic.

### 2.  PROGRAM STORAGE

### RAM

Random Access Memory is semiconductor memory used for storing the central processor operating programs and data. RAM is read/write memory and is volatile (contents erased when power is lost). Battery powered RAM provides a small area of memory with a limited amount of protection should power be lost.

   MOS (metal oxide semiconductor) and FET (field effect transistor) devices are different types of semiconductor technology used for RAM.

### ROM

Read Only Memory is normally used for storing the central processor system software. It is a non-volatile semiconductor memory device (contents retained when power is lost).

   Software is written into ROM at the time of manufacture and cannot be changed afterwards. PROM (Programmable ROM) allows software to be changed once only after manufacture and

EPROM (Erasable PROM) allows unlimited changes to be made by the supplier.

## Bubble Memory

Bubble memory is a non-volatile read/write memory device which, when combined with some RAM, offers an alternative main/back-up memory facility to that of RAM/ROM and magnetic tape/disk.

## 3.  POWER SUPPLIES

Most PABXs require a 50 volt dc power supply. This is normally achieved by mains rectification with battery back-up to provide service for a pre-determined time (typically 1 to 4 hours) in the event of mains failure. If standby generators are available, then battery capacity is reduced to a size sufficient only to provide power during generator start-up.

With no standby generator available, a mains outage greater than the battery reserve will result in an eventual total loss of service and possible erasure of operating software (programs) from the control system. These programs will need to be reloaded into the central processor when power is restored. With many PABXs, this is done automatically.